Expanding Private Production of Defense Services

Frank Camm

NATIONAL DEFENSE RESEARCH INSTITUTE

RAND

Prepared for the Commission on Roles & Missions of The Armed Forces

PREFACE

This report discusses key issues involved in outsourcing Department of Defense (DoD) support activities. It outlines an approach to addressing persistent barriers to outsourcing and provides guidelines for pursuing an outsourcing policy. It should be of interest to analysts and policymakers studying outsourcing for DoD.

The work is part of a larger project undertaken by RAND at the request of the Commission on Roles and Missions of the Armed Forces, which sponsored the research. The Commission was created in 1993 by Congress to review and evaluate "current allocations among the Armed Forces of roles, missions, and functions" and to "make recommendations for changes in the current definition and distribution of those roles, missions, and functions" (National Defense Authorization Act for FY 1994, *Conference Report*, p. 198).

The objective of RAND's overall effort has been to provide analytic support to the Commission's deliberations. The views expressed in this report are those of the author. The Commission does not necessarily endorse the options presented, the methodology involved, or the discussion contained in this report. This report represents one of many inputs provided to inform the deliberations of the Commissioners, who applied their own experience and judgment in arriving at the conclusions and recommendations that are found in the Commission's final report, *Directions for Defense.*

The research was conducted within the National Security Research Division of RAND's National Defense Research Institute (NDRI).

NDRI is a federally funded research and development center sponsored by the Office of the Secretary of Defense, the Joint Staff, and the defense agencies.

CONTENTS

FIGURES

SUMMARY

Among other topics, the Commission on Roles and Missions of the Armed Forces considered whether the Department of Defense (DoD) should contract out—*outsource*—support services that DoD now produces in-house and, if so, (1) which services DoD should outsource and (2) what DoD can do to make outsourcing more cost-effective. This report reviews a set of issues relevant to these questions. The report first reviews barriers and objections to outsourcing raised by earlier studies and government commissions that have addressed outsourcing. It then reviews insights from commercial-sector experience with outsourcing that DoD could use to guide its own actions on outsourcing. Finally, it offers suggestions about how to structure an implementation plan for large-scale outsourcing of support services. In particular, it identifies the attributes of support activities that DoD should consider outsourcing first and how DoD could facilitate an outsourcing program.

ADDRESS BARRIERS AND OBJECTIONS

The first step toward expanded outsourcing is to acknowledge those objections to outsourcing that persist and, wherever possible, develop outsourcing strategies explicitly designed to overcome them. Those objections fall into three broad categories: (1) economic injury to political constituents, (2) increased fraud and abuse by contractors and government officials, and (3) failure by contractors to provide support services needed during combat.

Key decisionmakers worry that expanded outsourcing will hurt their constituents. Evidence from the Base Realignment and Closure

(BRAC) experience suggests that negative effects will be smaller than most people expect, but additional outsourcing will hurt some people as it moves the location of federal spending. DoD has a variety of options for mitigating such injury, some of which the Commission has proposed.

Key decisionmakers who associate contracting for defense services with fraud and abuse fear that more contracting will inevitably lead to more fraud and abuse. Such fear obscures the fact that most defense contractors perform well. The most serious problems stem from a few bad actors. DoD should be able to reduce fraud and abuse by giving greater attention to the reputations of potential sources during source selection. But fraud and abuse will not go away. When it is detected, the government should react not by adding contracting regulations that seek to eliminate such abuse in the future but by punishing the offenders severely enough to deter such behavior in the future.

Finally, key decisionmakers fear that contractors will not provide the support services required during a contingency. Again, most contractors have performed well during contingencies. But the potential for misunderstandings or even misfeasance rises as a military commander requires more real-time control over a support service, relies more heavily on an untested commercial source for surge capability, and requires the support service in a more hazardous theater of operations. More-thorough integration of a contractor into military planning and execution during peacetime and more-exhaustive planning for contractor operations other than normal peacetime operations can help DoD reduce all these problems. Until such integration and planning can be achieved, however, military commanders should be cautious about their willingness to rely on support services that directly affect their ability to operate during a contingency.

CHOOSE THE BEST SOURCES OF SUPPLY THAT AVAILABLE CONTRACTING VEHICLES ALLOW

The second step toward expanded outsourcing is to identify the activities that are most cost-effective to outsource. A review of the empirical literature on commercial outsourcing suggests that successful commercial firms become more reluctant to outsource an activity as (1) real-time control of a complex process becomes more important, especially in an uncertain operating environment; (2) the potential joint value to buyer and seller of employing customized assets grows, especially in an uncertain operating environment; or (3) it becomes harder to specify the performance desired in a contract well enough to enforce the contract in court.

Successful commercial firms are more likely to split any particular workload between internal and external sources when the following circumstances apply: (1) an external source looks more cost-effective, but an internal source provides a setting for training personnel who will oversee the external source as a yardstick against which to compare the performance of the external source; (2) an internal source looks more cost-effective, but an external source provides a source of market and technological information that the buyer can acquire in no other way; or (3) having an internal source and an external source directly compete against one another on a continuing basis increases the competition that each faces, thereby enabling the buyer to extract better performance from both.

Although DoD's institutional setting differs from that of most successful commercial firms, the same factors that promote the decisions in the commercial sector discussed in the preceding paragraph relate to decisionmaking in DoD as well. Innovations in commercial contracting practices over the past 15 years have encouraged commercial firms to rely more heavily on external sources of support services by making it easier for these firms to overcome concerns like those raised above. DoD should recognize that its ability to outsource cost-effectively depends heavily on the contracting vehicles that it can use. Current DoD contracting practice severely limits DoD's ability to follow the commercial move toward increased outsourcing. Contracting reform could help DoD overcome a number of important barriers to expanded outsourcing.

STRUCTURE AN IMPLEMENTATION

The two steps above will succeed only if DoD can devise a long-term program of organizational change. That is, to expand outsourcing in a way that gives DoD full access to the core competencies for support services being displayed in the commercial sector, DoD will have to change many of the processes it uses to plan for and execute support services. To change all these processes, DoD will have to accept a degree of cultural change that most large organizations find very hard to accept.

DoD is more likely to succeed with significant change if it develops an effective, long-term implementation plan for phased expansion of outsourcing. The plan starts at the top, with clear and sustained support from the senior leadership. It includes the identification of a "champion" empowered to work across functions and organizations in DoD and to sustain the intent of senior management as implementation moves from defining policy to changing specific practices at depots and installations. It includes a mechanism for ranking activities in terms of the likelihood that outsourcing them would improve the cost-effectiveness of DoD's support services without imposing any unacceptable risks should outsourcing not work as anticipated. And it includes a plan to outsource the most-promising activities first. The plan should give close attention to achieving a series of early successes that build confidence in expanded outsourcing within DoD and enable DoD to outsource more and more-challenging services as its confidence and capability grow over time.

Any outsourcing decision should give close attention to available contracting vehicles that would give DoD access to an external source. The long-term implementation plan should promote contracting reform and build on it, incrementally expanding outsourcing as better contracting vehicles become available.

RECOMMENDATIONS

This report makes four basic recommendations.

1. **Plan formally for the major organization changes required to implement and sustain privatization.** Privatization involves far more than a simple shift from an in-house source of supply to a contract source of supply. Successful privatization on the scale that DoD is now discussing will require significant changes in many DoD processes. A number of these changes are cultural; they will be far harder to effect and diffuse than just making simple changes in legislation, regulations, or formal policies. For these changes to succeed,

DoD needs to develop a workable implementation plan and stick with it over the period of time it will take to achieve these changes.

2. **Focus on improving contracts.** By definition, DoD cannot expand privatization without increasing its dependence on contracts. Effective contractual vehicles are the key to getting the performance that DoD needs from new contractor sources of supply. Keeping this fact in mind, DoD should carefully coordinate any privatization effort with its efforts to reform acquisition, or it should initiate new efforts to improve the contractual vehicles it will use to access contractor sources of supply.

3. **Start with the best candidates for privatization.** DoD should seek early successes in expanded privatization and use those successes to increase confidence in the Congress and throughout DoD that still further privatization is promising. Conversely, it should assiduously avoid early failures that could discredit the longer-term program of privatization. Both considerations suggest that DoD should start where expanded privatization is easiest to achieve and move—over time, as contractual improvements and accumulated experience build DoD's capabilities—to privatize ever-more-challenging activities.

4. **Where necessary, protect constituents that privatization might hurt.** DoD should recognize that not everyone will benefit from privatization, even if it is outstandingly beneficial for DoD as a whole. DoD should expect political opposition and seek constructive ways to address it. One approach would seek simple ways to limit the losses induced by privatization. Another would use an extrapolitical process, such as the current BRAC process, to choose a group of candidates for privatization, thereby diffusing the losses caused by privatization.

ACKNOWLEDGMENTS

More than most RAND reports, this report brings together the ideas of many people. I especially appreciate the analytic and policy insights of my RAND colleagues John Bondanella, Mary Chenoweth, David Chu, Carl Dahlman, J. D. Eveland, Donna Fossum, Ken Girardini, Don Henry, Jeff Luck, Elan Melamid, Dean Millot, Nancy Moore, Ray Pyles, Sue Resetar, Ken Reynolds, and Hy Shulman, and of Michael Hovey and Gene Porter, members of the staff of the Commission on Roles and Missions. I have had the opportunity to present earlier versions of this material to many current and former senior officials in the Army, Navy, Air Force, and Office of the Secretary of Defense, and to benefit from their candid and constructive reactions. Dave Adamson, Marian Branch, and Paul Steinberg substantially improved the presentation of the material. Finally, David Chu deserves special thanks for supporting this work from its inception through a long series of high-level presentations arranged through his efforts. I lean heavily on the contributions of everyone mentioned above, but in the end, of course, I accept full responsibility for the final product.

Chapter One

INTRODUCTION[1]

The fall of communism, by definition, has precipitated privatization of publicly owned enterprises throughout Russia and Eastern Europe. This change has reignited the intellectual and ideological debate of the 1930s over the relative efficacy of privately and publicly owned enterprises. Privatization is now in vogue in the West as well. "Privatizing" (there is confusion about this word, discussed later in the report) Department of Defense (DoD) support functions is a variation on this theme. Discussions in the Commission on Roles and Missions of the Armed Forces (CRMAF) pointed to this global shift to the private sector as one more reason to reopen a very old debate about contracting for defense services.

The ongoing global revolution in commercial business practices is encouraging organizations to "outsource" much of what they used to do in-house and to focus their in-house activities on the things that,

[1]In the interest of brevity and simplicity, I do not identify the specific sources for all the arguments offered here. Throughout the report, I draw heavily on insights from Alfred D. Chandler, *Strategy and Structure*, MIT Press, Cambridge, MA, 1962; Kathryn R. Harrigan, *Strategies for Vertical Integration*, Lexington Books, Lexington, MA, 1983; Steven Kelman, *Procurement and Public Management*, AEI Press, Washington, DC, 1990; Donald Kettl, *Sharing Power*, Brookings Institution, Washington, DC, 1993; Thomas McNaugher, *New Weapons, Old Politics*, Brookings Institution, Washington, DC, 1989; Ian R. Macneil, *The New Social Contract: An Inquiry into Modern Contractual Relations*, Yale University Press, New Haven, CN, 1980; Michael Smitka, *Competitive Ties*, Columbia University Press, New York, 1991; Oliver E. Williamson, *The Economic Institutions of Capitalism*, Free Press, New York, 1985; and James Q. Wilson, *Bureaucracy*, Basic Books, New York, 1989. A similar and complementary view of outsourcing appears in Fred Thompson and L. R. Jones, *Reinventing the Pentagon: How the New Public Management Can Bring Institutional Renewal*, Jossey-Bass, San Francisco, CA, 1994, pp. 157–162.

strategically, matter to them most—their "core competencies."[2] Taken together, these trends help support a growing enthusiasm for outsourcing many DoD support activities to the private sector. Such outsourcing would presumably help DoD to focus in-house on its "core" military concerns and harness the power of private enterprise to provide more cost-effective support services.[3]

The final report of the CRMAF gives a central role to expanded outsourcing of support services.[4] Expanded outsourcing of support services is not a new idea. Motivated by a belief that commercial firms offer higher-quality and lower-cost products and more-rapid innovation than comparable government agencies, past commissions have recommended expanded outsourcing repeatedly over the last few decades.

The current Commission supports this recommendation for similar reasons. But it also recognizes the importance to DoD of the revolution in business practices under way in the commercial sector today. Firms are rapidly learning to apply information technologies in new ways to increase their responsiveness to customers and at the same time reduce their costs. The Commission believes that DoD can get direct access to the products of this revolution by contracting for a larger fraction of its support functions. More subtly, the Commission anticipates that DoD can emulate successful commercial firms by outsourcing noncore activities now conducted in-house. If the Commission is correct, expanded outsourcing of support activities

---

[2]"Outsourcing" is the practice of buying a product—a good or a service—from an outside source rather than producing it in-house. "Core competency" has as many definitions as there are commentators who use the term. We use it here in a very broad sense to denote a capability that an organization must retain in-house to ensure its survival and long-term success. Such capabilities tend to define the *raison d'être* of the organizations where they reside—organizations that would not be what they are if these core competencies did not exist.

[3]The DoD view of "core" should not be confused with "core competency"; the two terms have evolved independently and, at least in principle, could be quite different. However, both reflect strong beliefs about activities so important that they cannot be outsourced. Over the years, DoD has defined such *core activities* in many different ways. Note the strong assumptions implicit in the argument in the text that DoD's core is related to core competency and that firms in the private sector can provide support services more cost-effectively than DoD.

[4]Commission on Roles and Missions of the Armed Forces, *Directions for Defense*, Washington, DC, May 24, 1995, especially Chapter 3.

will help DoD adjust to a falling budget and a new world environment without giving up the quality of support that it needs to ensure an adequate national defense.

The traditional DoD approach to outsourcing proceeds in two steps (see Figure 1.1). The approach first distinguishes "core" from "noncore" activities; "core" activities, by definition, cannot be outsourced. It then divides noncore activities into those that DoD will produce internally and those that it will procure from an external source. The Commission's approach can be seen as a new version of this approach. It defines *core* narrowly to include only those activities that are "inherently governmental"—activities that, for legal or even constitutional reasons, it would be inappropriate to relinquish responsibility for to an outside source. The Commission then uses a presumption of superior private-sector performance to prefer a private to a public (governmental) source for noncore activities. The burden of proof lies on those who prefer a public source.

Three aspects of this approach give us pause:

(1) The simplified presumption in favor of a private source limits any effort to weigh the costs and benefits of public and private sources for any particular support service. As long as an activity is not inher-

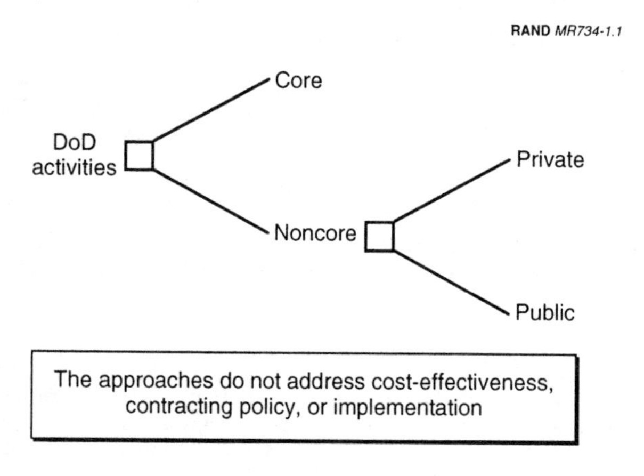

Figure 1.1—Common Approaches to Outsourcing Follow a Simple Decision Tree

ently governmental, we presumably want to have it produced in the most cost-effective manner possible.[5]

(2) The approach gives limited attention to the difficulties that must be overcome to maintain an effective contractual relationship with a private-sector source. DoD cannot gain access to any private source of support without using a formal contractual arrangement.

In contrast, this report explicitly uses cost-effectiveness as a basis for asking which DoD support services should be outsourced. To do so, it makes no prior judgment about the inherent governmental nature of an activity. And it considers attributes of both the source of a support service and the "governance" structure that any DoD activity buying this service uses to get access to it: a direct command-and-control link when the DoD buyer "owns" the seller within an armed service; a memorandum of agreement when the buyer and seller lie in different parts of DoD; or a government contract when the seller is a private firm.

From this perspective, outsourcing is a choice between one imperfect "bundle"—private-production arrangements and a government contract—and many other imperfect bundles of government production arrangements and more-or-less formal intergovernmental agreements. Today, the government-contracting mechanism can significantly limit communication and control between buyer and seller. It can impose substantial administrative costs on buyer and seller, who must observe complex regulatory requirements and maintain complex accounts. It can also induce government contractors to behave quite differently than the commercial firms that typically come to mind as characteristic of entrepreneurial enterprise. Which bundle offers the most effective unity of purpose between DoD and its suppliers at a reasonable cost? *Whether a privatized or*

[5]The term "cost-effective" means different things to different people. Here, "cost-effectiveness" requires that two conditions be met. (1) We produce any level, quality, or mix of support services at the lowest *social cost*—what people or firms are willing to pay for the resources absorbed by DoD in the resources' next best use—possible. And, given this first condition, (2) we produce the level, quality, and mix of support services that maximize the difference between the total social benefits and costs associated with these services. "Cost-effectiveness" specifically does not mean that we minimize the cost of producing a service at the lowest level of quality acceptable.

government bundle dominates is an empirical question, likely to differ from one case to the next.

(3) The third aspect of the Commission approach that concerns us is the limited attention it gives to factors that should be considered to ensure successful implementation of any proposed outsourcing. To the contrary, the Commission implicitly promotes a rapid program of outsourcing services that could lead to early failures. That is, if DoD pursues extensive, expanded outsourcing without giving such factors adequate attention, it could fail to realize its expectations about improved performance and reduced costs. Such failures could discredit the notion of expanded outsourcing before such outsourcing has a chance to prove itself.

To support the Commission, we developed a broader framework for weighing the pros and cons of outsourcing any particular support service that DoD currently produces in-house. For the purposes of this report, we present this framework as a series of screens (see Figure 1.2):

1. Would outsourcing be acceptable to key decisionmakers?
2. Would outsourcing be cost-effective? We found that we could not address this question without asking what contractual vehicles were available to allow outsourcing.
3. Could DoD implement the proposed outsourcing?

There is nothing sacred about this particular sequence.[6] We use it simply to impose some order on a complex problem. The following chapters of this report consider each of these questions in turn.

Before we turn to these questions, however, it is worth highlighting a perspective that will appear repeatedly throughout the report. Dr. Steven Kelman, currently director of the Office of Federal Procurement Policy in the White House, has developed a cogent argument that maintains that almost all facets of federal procurement

---

[6]In fact, ongoing work at RAND to put the analysis reported here into operation suggests that DoD will want to consider factors relevant to outsourcing decisions in a somewhat different sequence. Details on that sequence await further analysis and testing.

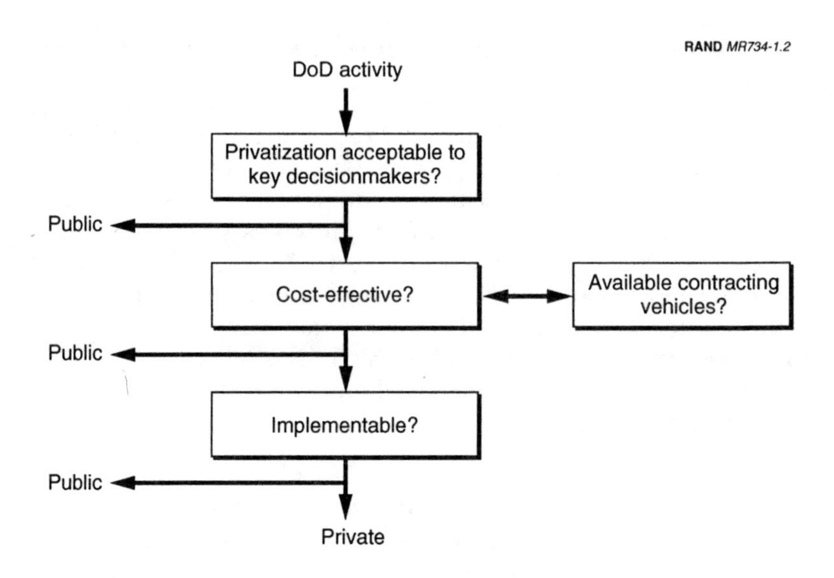

Figure 1.2—An Alternative Approach Addresses Outsourcing Issues Through a Series of Screens

law, regulation, and practice can be traced back to one of three core values shown in Figure 1.3.[7]

1. To ensure *equity*, the federal government wants to ensure equal access to government contracts for all qualified suppliers—"free and open competition." To cast the net as broadly as possible, qualifications are kept to the minimum-acceptable level. The government must give private parties who feel that they have been treated unfairly the right to protest that treatment in an open and accessible manner.

2. To ensure *integrity*, the federal government pursues a "zero tolerance" approach to fraud and abuse. As a matter of principle, it will not tolerate any level of fraud by government employees or contractors. To ensure full accountability, it insists that the whole

---

[7]Kelman, 1990.

RAND *MR734-1.3*

Equity	• Ensure equal access for *all* qualified suppliers
Integrity	• Prevent *all* government-employee and -contractor fraud
Efficiency	• Get *lowest cost* for acceptable performance

Outsourcing in the commercial sector is not compatible with these values

Figure 1.3—Current Defense Contracting Policy Is Based on Strongly Held Values

process of contracting-related decisionmaking be open to public scrutiny and potential challenge.

3. To ensure *efficiency*, the federal government typically seeks to procure any product at the lowest cost possible to get an acceptable level of performance, which is typically kept low to maintain the focus on cost.

These values have a powerful hold on the contracting culture in DoD. As reasonable as they might sound to many of us, they differ from the values that increasingly dominate successful commercial firms in their outsourcing activities:

1. Commercial firms increasingly raise their standards for qualified suppliers, limiting access for suppliers. These firms then deepen their relationships with the suppliers that remain, further limiting the ability of other suppliers to offer them services.

2. Commercial firms experience fraud and abuse in their outsourcing activities. But increasingly, they are learning to accept such problems when they cost more to eliminate than to live with.

3. Commercial firms increasingly emphasize performance over cost, giving increasing attention to subtleties of performance that may

> be difficult to justify objectively.[8] They are finding that such an approach ultimately yields higher cost-effectiveness, even if they end up paying more for the products they buy.

This alternative perspective will be problematic, if not impossible, for the federal government to adopt. Therefore, DoD officials should recognize that it may be difficult for the government to emulate all of the contracting practices that have facilitated cost-effective outsourcing in the commercial sector. Recognizing the limits of government contracting, DoD officials should be prepared to deal with these limitations effectively as they prepare to expand outsourcing dramatically. The following chapters offer a variety of illustrations of this point.

---

[8]Cutting *cycle time*—the time between a customer's order and its fulfillment—can be monitored and justified objectively. Investing in anticipation of a customer's future needs is much harder to monitor or justify before the need becomes apparent. Despite this fact, successful commercial firms increasingly seek suppliers who can "grow with them" precisely by making such investments.

Chapter Two

CONCERNS OF KEY DECISIONMAKERS

Past commissions have identified three major factors that decisionmakers worry about when they consider expanded outsourcing:

- Political leaders worry that expanded outsourcing will transfer jobs and the cash flow associated with them from communities where DoD currently produces organic support services.
- Political leaders also worry about the fraud and abuse so often linked to defense contractors in the public forum; expanded outsourcing could well lead to more fraud and abuse as contractors play a larger role in providing support services.
- Combat commanders worry that private sources of support will not provide predictable services when they are needed during a contingency.

This chapter discusses these factors, which have dampened support for expanded outsourcing in the past. It draws first on past commissions to identify important issues, then on RAND and other sources to identify policy options that could help DoD address these issues.[1]

---

[1]These commissions are discussed in unpublished RAND work by Scott Harris on effective advisory commissions.

CONCERNS ABOUT NEGATIVE ECONOMIC EFFECTS

Outsourcing could very well have negative economic consequences in localities where DoD currently produces in-house support services. DoD support activities are often large employers—sometimes the largest single-point employers in a state. Their size, singly and jointly, gives them great political significance, a fact reflected, for example, in the powerful congressional depot caucus that has organized around the districts in which defense depots reside.

To a first approximation, outsourcing would simply move jobs—and the cash flow associated with them—from the communities where they exist today to the communities where the relevant private-sector sources of supply exist. But the interest groups organized around existing activities are typically better organized and motivated than those associated with potential new activities. Outsourcing could easily diffuse jobs and cash flow from a few large federal facilities to many smaller private-sector locations having less political visibility. Because the lobbyists associated with original equipment manufacturers that might produce support services are only now shifting their focus from the acquisition of new systems to the support of existing systems, they do not present an effective political counterweight to the depot caucus. And in the long run, outsourcing is, in fact, likely to be associated with a reduction in jobs and the cash flow associated with them. So the politics of defense support services are loaded against radical change.

The following are pressures that favor the status quo. Along with them we present options for overcoming them.

(1) Any outsourcing effort that displaces government workers could threaten their pensions. The Office of Personnel Management is limited in its ability to craft specially targeted programs to protect the pensions of displaced workers. But programs to allow early retirement or cash-outs deserve greater attention.

(2) DoD could favor commercial sources near current government workers. Many services, such as base operating support and many morale-welfare-recreation (MWR) activities, must be produced close to the place where they are consumed. In these circumstances, DoD

will automatically prefer local private sources. Current Air Force efforts to privatize depot operations "in place" at Newark, Ohio, and, now, at Sacramento, California, and San Antonio, Texas, represent efforts to favor private sources near the existing source of supply. Most logistics activities, however, could be produced at the DoD site or elsewhere. For example, DoD routinely ships electronic components to contractor facilities throughout the country to be maintained. Information-management services can as easily occur at a regional commercial center as at a DoD site. Conversion of government operations to government-owned, contractor-operated (GOCO) facilities is another option. More generally, DoD could favor local sources in the criteria it uses for source selection. Or DoD could offer facilities and other assets at existing locations to contractors at favorable enough terms to induce them to use those assets, thereby keeping the work close to home.

(3) DoD could favor organic sources for systems now maintained in-house, in recognition of the political difficulty of moving such work to an alternative source. The Commission implicitly grandfathers current depot-level support by differentiating new weapon systems from existing weapon systems. It recommends that DoD plan to use only contractor logistics support for all new weapon systems. The Commission is more cautious about outsourcing depot-level support for existing weapon systems, especially when the government does not own the technical data required to hold an effective competition and the original equipment manufacturer is therefore the only realistic private source of the services. Similarly, because commercial firms have shown only limited interest in supporting DoD systems that use outdated technology, few commercial sources of such services are available, and those that are need not be especially responsive to DoD. DoD could consider grandfathering organic support services for such aged systems.

(4) The experience of the BRAC Commission offers three insights for expanded outsourcing.

First, economic studies by the OSD Office of Economic Adjustment have shown repeatedly that the negative economic effects of closing a DoD installation are normally short-lived.[2] Within two years, the number of jobs in most communities equals or exceeds the level before closing. A number of communities have seen substantial improvements as redevelopment takes hold. Many individuals do experience lower wages, pension losses, and lower real-estate values. But on net, the long-term story in most communities that have lost DoD installations is positive. Large-scale outsourcing that led to high levels of job loss would likely display similar long-term effects. If these results can be documented and presented in clear terms that relate to the communities at risk, political opposition to outsourcing should ease.

Second, the experience with BRAC to date suggests that it has been easier to shut down existing DoD activities in locations where the facilities have higher value in alternative uses and the workers have many alternative employment opportunities. These tend to be locations where DoD employment does not dominate the local labor market, and where the local economy is fairly diversified and is growing. Outsourcing of support services is likely to be politically easier in such places as well.

Third, a broad approach to outsourcing like that taken by the BRAC Commission—an approach that spreads the political pain associated with eliminating government jobs in a way that overcomes political opposition to the change—may be necessary to achieve the degree of outsourcing contemplated by the Roles and Missions Commission. We have found broad interest for such an approach in DoD. However, the approach has a downside: Specific support services are likely to be linked more directly to specific military capabilities than specific bases have been in the BRAC proceedings. Although important organic services at a base closed by a BRAC decision could always move elsewhere, DoD should be careful about relinquishing to an outside commission its control over how best to integrate support services with its military capabilities.

---

[2]Personal communication, July 1995, with Carl Dahlman, who oversaw many of these studies while an official in OSD.

CONCERNS ABOUT FRAUD AND ABUSE ASSOCIATED WITH CONTRACTING

Abuse is unusual among DoD contractors but is highly visible when it comes to light.[3] Many anecdotes exist about firms exploiting their monopoly positions during the Gulf War, firms that were qualified during the 1990s' experiments with public-private competition to increase the number of competitors participating but that had no plausible capability to perform the work involved, and firms that have explicitly cheated DoD in a variety of ways. However, at least as many anecdotes exist about firms that have performed beyond the literal language of their contracts when called upon. We can expect this pattern to continue. But expanded outsourcing will be easier if DoD can find a better way to react to fraud and abuse when it occurs.

Today, Congress generally reacts to new evidence of abuse by changing the law to ensure that each new type of abuse encountered will not occur again. The result is a cumbersome body of law and regulations that raises the cost of dealing with the government and, hence, reduces the field of commercial firms willing to provide services to DoD. The commercial sector offers two alternatives.

First, commercial firms increasingly rely on reputation as a basis for choosing suppliers. DoD is beginning tentative experiments with a similar approach. The Air Force, for example, plans to use its contractor performance assessment reports (CPARs) more extensively as an objective source of information that future source selections can draw on to distinguish suppliers having different past histories with DoD.[4] Great skepticism remains about whether DoD will succeed in

---

[3]*Abuse* means different things to different people. We give it a broad meaning here, encompassing not just legal violations but more-general contractor failures to perform as promised and instances in which contractors exploit their special relationship with DoD to extract revenues that exceed the levels they could get in the presence of meaningful competition. The focus here on contractor abuse reflects our approach: We are reviewing the concerns of key government decisionmakers. Congress and DoD have not been immune to failing to follow through on their own promises or unfairly exploiting their own special relationships with DoD contractors.

[4]Some parts of the Air Force have relied on the CPARs for many years, in both source selections and less-formal settings. One senior official, for example, would not meet with a contractor without first reviewing its CPAR file to provide a context for the discussion.

leaning more heavily on assessments of reputation in its source selections. The following types of concerns arise:

- When DoD deals with a large firm, which part of the firm should it examine to determine the "reputation" of the particular division making an offer?
- If DoD rejects an offer on the basis of a firm's reputation, what can that firm do to rehabilitate itself?
- How can DoD judge the reputation of firms it has not dealt with in the past?
- What standards will emerge to ensure that any government assessment of reputation is objective?

The courts will have to develop answers to these questions as rejected offerors challenge the results of source selections, and that process will take time. But major changes in government rules affecting the private sector typically require such a transition period.

Second, firms are increasingly learning to avoid spending more to eliminate abuse than it costs to tolerate it. Much anecdotal evidence points to examples where the government, in its efforts to eliminate fraud and abuse, spends more than it recovers. DoD should recognize that no set of procedures can avoid all fraud and abuse. And tolerating the existence of fraud does not mean treating it lightly when it is detected. When individual incidents occur, the response to them should not be to revisit the procurement regulations but to punish the perpetrator heavily enough to provide a deterrent for others in the future. That is, enforcement resources should focus on the isolated wrongdoers when they are caught and not on the activity of contracting as a whole. This is a strategy that commercial firms are continuing to refine.

Those who hold the federal government to a higher moral standard on such issues will make it hard for DoD to follow commercial practice. If Congress, in particular, perceives that DoD is unwilling to enforce the rules in place, it may decide to tighten the rules rather than to tighten enforcement of existing rules. And scandals offer too easy an opportunity for those seeking an opening to exploit them for whatever end they have in mind. In the end, for this approach to work, Congress must stand behind it.

That said, in practice, DoD is already taking steps in this direction. Internal management rules for initiating actions against alleged abuse direct government resources to the largest sources of abuse. And innovations, such as allowing government employees to use credit cards to make small purchases for government use without careful review of each purchase, recognize the value of greater discretion, even if it potentially enables some additional abuse.

CONCERNS ABOUT PREDICTABLE SUPPORT DURING A CONTINGENCY

The very phrase most often used to denote relationships inside organizational hierarchies—command and control—comes to us from the battlefield, where commanders understand the importance of unity of purpose. Can outside contractors support a combat commander's needs effectively: providing real-time responsiveness, providing surge capability, and providing people and other valuable assets in a combat zone? Or are formal command-and-control relationships within DoD necessary in this setting? These questions have provided the basis for three basic concerns about outsourcing in past discussions.

Real-Time Control

No one faces a more complex management environment than a military commander in wartime. War occurs at the end of long, uncertain support links, often in an unfamiliar, unprepared theater. Information on the status of forces is limited at any point in time, and that status changes quickly without warning. This complexity has always presented a challenge. The advent of continuous, 24-hour-a-day execution and short decision cycles under combat concepts such as AirLand Operations, combined with greater uncertainty than ever about what form the next contingency will take until it arises, make the management of forces in wartime more complex than ever. Combat commanders can succeed in such an environment only with well-trained forces that, repeatedly during peacetime, have simulated the complexity they will face during war. These forces must be able to act with unity of purpose, despite all the confusion that accompanies war.

Real-time control is important to commercial firms facing complex environments as well. The first large American commercial businesses—the railroads—arose from a perceived need to impose close coordination on a complex process in the presence of limited information about the status of resources. These firms found it impossible to use contracts to create the unity of purpose they needed for controlling their complex processes. As the complexity and interdependence of industrial processes rose, other large firms learned the same lesson. Should DoD, which faces far more demanding circumstances during warfare, attempt to face them by contracting for support services critical to maintaining unity of purpose? DoD decisionmakers should consider two points.

First, when they have a choice and when real-time control is critical, DoD buyers should be cautious about relying on contractors (or on DoD suppliers too organizationally distant from them within DoD). In the world of defense support services, real-time control is most critical near combat—in organizational and intermediate support activities. The more loosely coupled a support activity is to ongoing combat activities, the less important this concern becomes. Where to draw the line is an inherently empirical matter, likely to differ from one situation or support activity to the next.

Second, contractors have historically provided support services critical to the real-time success of combat. Contracting for critical support services was routine in the nineteenth century; in this century, it has been common since at least the Vietnam War.

DoD routinely depends on contractors to support the weapon systems they developed while it continues to develop the technical orders, support equipment, and training programs that will ultimately allow DoD personnel to assume responsibility for support. DoD often relies on such interim contractor support for a decade or more following initial operational capability for a new system. In recent years, DoD has shown no strong desire to escape the grasp of contractors by accelerating the transition from interim contractor support to organic support. That is, DoD regularly and comfortably relies on contractors for the support required by its most advanced weapon systems. In general, this approach has worked well.

DoD has also routinely relied on contractors for many other services in-theater during contingencies. The Army even has a formal program called the Logistics Civil Augmentation Program (LOGCAP), which it uses to procure contractor-provided base operating support in-theater during contingencies. This program has been active in every military action in which the Army has participated during the last two decades. Although it has reservations about whether it is paying too much for this service, the Army is pleased with the quality of support it receives through its LOGCAP.

Even though the DoD experience with contractors in-theater has generally been good, bad things do happen. By definition, the uncertainty that pervades hostilities creates surprises. DoD and its contractors can easily interpret these surprises differently, which leads to serious disagreements about what is expected on each side. And contractors do fraudulently exploit such surprises on occasion. Everyone has anecdotes about genuine bad actors during contingencies. DoD would benefit from systematic empirical assessment of such experiences to identify (1) the factors that contribute to integrating contractors successfully into combat operations, and (2) the ways in which contractors, so integrated, and analogous organic sources of support compare in their actual, demonstrated abilities to respond to combat commanders' needs in real-time.

A factor likely to be important to effective unity of purpose in wartime is systematic integration of contractors into operations during peacetime. Only by building trust in realistic situations during peacetime can the military and contractors be comfortable enough so that they can anticipate one another's responses sufficiently to be able to rely on one another during wartime.

Building such mutual trust takes a long time and a lot of time—one of the unanticipated lessons of recent commercial efforts to expand outsourcing through "agile" strategic alliances with other firms.[5] To build such trust across contractual boundaries, successful commercial firms are exploiting new information technologies and are founding integrated planning and execution processes on that trust

---

[5]Advocates of "agile" firms expect them to move nimbly in and out of relationships to exploit business opportunities as they arise. Most successful strategic alliances between firms, however, build on years of mutual testing and adjustment.

over the longer term. Outsourcing becomes more attractive in commercial business as such trust and the integration it allows grow.[6] DoD should expect to benefit from improved contracting arrangements that enhance its effective control over its suppliers. Unfortunately, current procurement regulations often discourage the long-term relationships required to build mutual trust.

Surge Capability

It is hard to judge the cost-effectiveness of surge capability during peacetime, unless the assets that will provide this capability are carefully segregated from other assets and tested periodically to ensure their readiness within an acceptable decision horizon.

Within DoD, support-service planners and resource managers typically do not explicitly distinguish between resources dedicated to peacetime workload and those maintained only to ensure surge capability. Planners can often use operations research tools to simulate workloads anticipated in various contingencies and the resources that will be needed to support those contingencies. Such studies are often considered in resource planning and funding decisions for various support activities. But in the end, the planners generally create support activities that can meet normal workloads in peacetime and use resources that exist in peacetime more intensively to meet surge requirements during a contingency. An outsourcing decision, particularly one focused on peacetime costs and performance, could easily view the slack now used to preserve surge capability as inefficient excess capacity.

An obvious alternative to this approach would explicitly separate peacetime and surge requirements and provide for them separately. Contractors could be paid to maintain certain specified assets ready for use in a wartime surge. In practice, such maintenance has proven to be extremely difficult. For example, requirements that petroleum

---

[6]Some management consultants speak of an extreme form of highly flexible outsourcing as a "virtual corporation"—an organization whose core competency lies primarily in coordinating its customers with its suppliers and includes no other in-house activity. This creature remains more an ideal concept than a reality. The very few major firms that have even approached virtuality remain more ongoing experiments in new management practices than practical, thoroughly tested exemplars for DoD.

firms maintain ready assets to clean up spills have not ensured that those assets were in fact ready when needed; the Valdez oil spill in Alaska illustrates this problem. An unusual actual attempt in the early 1980s to draw down the Strategic Petroleum Reserve in Texas in a peacetime exercise failed utterly. Owners of CRAF (Civil Reserve Air Fleet) aircraft, whom DoD had subsidized for years in exchange for making these aircraft available when called, lobbied to discourage DoD from asking for these assets during the Gulf War.

Three problems related to surge capability can be derived from the preceding discussion.[7]

First, it is hard for the government or a contract provider to judge a potential capability until that capability is tested. Organizations tend to focus on what has to get done *now;* an organization can easily lose track of the importance of ensuring that surge capability will work as planned until surge capability is actually needed. Because organizations tend to rely heavily on standard operating procedures and personal relationships tested in actual practice, capabilities not included in these procedures and relationships easily fall through the cracks.

Second, government and privately owned organizations benefit from practice and learning. If they have to learn how to provide surge capability and integrate it with other existing capabilities only when the surge occurs, they cannot provide the quality of service they could if they had prepared to surge.

Third, providing surge may not look nearly so attractive as a business opportunity when it is actually required as it did when the private provider contracted to provide it. Repeated tests have shown that otherwise apparently rational decisionmakers systematically underestimate the likelihood and cost of major, unusual events, such as a mobilization that would call upon contractor surge capability. Furthermore, up to the moment of mobilization, government payments to the private provider for surge capability are sunk. If the provider must use resources that have a high-value, alternative use—for example, cargo aircraft—to provide surge capability, it will resist the government claim on those resources unless the government

---

[7]See, for example, James G. March, *A Primer on Decision Making: How Decisions Happen*, Free Press, New York, 1994.

pays it enough during the surge itself to offset lost revenues from other sources.

The lesson is that DoD must be prepared to test the availability of the surge assets that it contracts for on a regular basis. One way to do so is to include these assets in normal peacetime training exercises. Such exercises will increase what DoD must pay for these assets—which is the price of ensuring that the assets are available when needed. And DoD should be prepared to pay for surge assets that have alternative uses when it needs them. A plan that shifted emphasis from preferred access to DoD business during the period before the Gulf War to increased payments during the war, when DoD needed assets, would have made it easier for aircraft owners participating in the CRAF program to withdraw their aircraft from commercial service.

Until DoD can trust that these assets will be available as needed, it should exercise caution about contracting for surge assets. As noted above, that trust may take many years of regular interaction to create and sustain.

The alternative to contract surge capability is organic capability in excess of normal peacetime requirements. In the absence of a formal process that forces DoD to consider alternative ways of providing support, organic capacity in excess of surge requirements can easily persist undetected. And, in fact, as the size of the total DoD force decreases and DoD weapon systems become more reliable, making fewer demands on the support system, such excess capacity is growing. DoD should be clear about which portion of that excess it continues to need as surge capability and which it will probably never need again. A reluctance to contract for surge capability should not become a willingness to tolerate excess organic capacity. In all likelihood, it will be easier to remove such excess from a contract source than from an organic source.

Production of Services in a Combat Zone

Concerns about wartime control and the availability of surge capability are present even if a contractor has no activities in the combat theater. Asking a contractor to produce services in-theater adds further concerns, which we address here.

As noted above, contractors have provided critical services in combat theaters for decades. But not all contractors are willing to do so. Facing the threat of Scud attacks, for example, one major defense contractor withdrew its personnel from the theater during the Gulf War. In prolonged negotiations, it would not accept any amount of money offered to remain; its commitment to its reputation for protecting its employees overrode any short-term focus on profits. Many other contractors, of course, remained throughout the war. Two issues must be focused on, therefore.

First, do contract employees have access to suitable protection and services in-theater? For example, even civilian DoD employees may carry life and health insurance that is invalid in a combat zone. At a minimum, DoD should ensure that all personnel supporting combat operations in-theater have access to reasonably priced insurance services. Evidence from the Gulf War suggests that civilian government and contract personnel did have adequate insurance coverage, but that great uncertainty prevailed about what coverage they actually had when the contingency arose. Relieving such uncertainty should make it easier for contractors to operate in-theater.

Contractor personnel should also get the same physical protection that analogous military forces receive. During the Gulf War, contractor personnel did not receive protection comparable to that received by DoD employees in-theater. For example, at the beginning of the contingency, contractor truck drivers carried troops who were fully equipped with gas masks forward, without gas masks for themselves. And contractor personnel who received military medical support in-theater found it summarily dropped following their evacuation from the theater if they were injured. Such differences can be, and should be, identified and removed to allow more-effective integration of contractors into the broader military effort in a contingency.

Problems that are more difficult to solve involve the legal status of contractors in-theater.[8] First, from a "friendly" perspective, to what extent are contractors subject to the Uniform Code of Military Justice? From an enemy perspective, are they combatants or civil-

---

[8]Raymond J. Sumser and Charles W. Hemingway, *The Emerging Importance of Civilian and Contractor Employees to Army Operations*, Landpower Essay No. 95-4, Association of the United States Army, Institute of Land Warfare, Arlington, VA, 1995.

ians? What difference would arming themselves for self-protection or wearing uniforms make? The changing political and technological nature of modern warfare strains the traditional answers to these questions currently reflected in international law. These issues require clarification that DoD, by itself, may not be able to provide.

Second, what advance understanding does DoD have with its contractors about their availability to provide services in a combat zone? Such an understanding is likely to include points raised in the first issue above but should also include a plan for integrating contractor operations into military operations during peacetime training and in particular contingencies. Current contracts for the maintenance of sophisticated defense electronics and engines often fail to specify explicitly the role of contractors in contingencies or expectations during contingencies. They often rely on boilerplate to spell out what is expected in unusual circumstances—with wording that is easy to disagree with and difficult to enforce effectively. Contracts supporting closer integration would focus greater attention on specific expectations during contingencies.

Contractors may be reluctant to execute plans when a war begins. Adequate plans should anticipate this problem and provide effective ways to confront it when the event arises. That is, even if workload does not change in a contingency, moving private assets into the theater has much in common with activating surge assets. Suddenly, the basic way a firm does business changes. All the concerns about planning for surge capacity apply here as well. Expectations about performance in-theater during a contingency should be critical parts of the performance that DoD says it wants when it chooses sources of supply and writes and monitors its contracts with them.

Taken together, these concerns point to the importance of better long-term integration of contract suppliers and military buyers to ensure unity of purpose during combat. This is the same approach that DoD routinely uses to manage misunderstandings between its operators (e.g., fighters, warriors, combat organizations) and its organic support organizations—careful integration of support activities into peacetime planning, training, and execution. Such integration requires long-term relationships with these contractors and enough mutual trust to ensure continuing, detailed discussions of requirements and capabilities. As DoD pursues such integration, it should not forget that it has a long history of satisfactory contractor participation in combat. DoD can learn valuable lessons from its own historical experience with contractors. That experience deserves more-careful empirical assessment.

SUMMARY

Key decisionmakers worry that expanded outsourcing will hurt their constituents. Evidence from the Base Realignment and Closure experience suggests that negative effects will be smaller than most people expect, but additional outsourcing will hurt some people because it moves the location of federal spending. DoD has a variety of options for mitigating such injury, some of which the Commission has proposed.

Key decisionmakers who associate contracting for defense services with fraud and abuse fear that more contracting will inevitably lead to more fraud and abuse. Such fear obscures the fact that most defense contractors perform well. The most serious problems stem from a few bad actors. DoD should be able to reduce fraud and abuse by giving greater attention to the reputations of potential sources during source selection. But fraud and abuse will not go away. When it is detected, the government should react not by adding contracting regulations that seek to eliminate such abuse in the future but by punishing the offenders severely enough to deter such behavior in the future.

Finally, key decisionmakers fear that contractors will not provide the support services required during a contingency. Again, most contractors have performed well during contingencies. But the potential for misunderstandings or even misfeasance rises as a military commander requires more real-time control over a support service, relies more heavily on an untested commercial source for surge capability, and requires the support service in a more hazardous theater of operations. More-thorough integration of a contractor into military planning and execution during peacetime and more-exhaustive planning for contractor operations other than normal peacetime operations can help DoD reduce all these problems. Until such integration and planning can be achieved, however, military commanders should be cautious about their willingness to rely on support services that directly affect their ability to operate during a contingency.

Chapter Three

COST-EFFECTIVENESS AND AVAILABLE CONTRACTING VEHICLES

Given that key decisionmakers believe that it is acceptable to outsource a particular activity, we now need to look at whether it is cost-effective to do so. We addressed this question by asking what factors successful commercial firms consider to decide which activities to outsource.[1] Outsourcing is the relevant commercial analog of privatizing support services now provided by organic organizations within DoD. Presumably, outsourcing occurs and persists in the commercial sector only when firms expect it to be cost-effective for them to do so.

In particular, we reviewed empirical studies of outsourcing decisions by successful firms since about 1945.[2] We found that willingness to outsource depends, in part, on the availability of suitable contracting vehicles. Recognizing that successful commercial firms have been using new contractual forms during the last 15 years or so, we gave special attention to how contracting considerations affect a firm's willingness to outsource.

Differences in the environments of DoD activities and those of analogous commercial firms can easily lead to differences in which

---

[1]This discussion draws heavily on Frank Camm, *DoD Should Maintain Both Organic and Contract Sources for Depot-Level Logistics Services,* RAND, IP-111, Santa Monica, CA, August 1993.

[2]Two good surveys of this literature are Paul L. Joskow, "Asset Specificity and the Structure of Vertical Relationships: Empirical Evidence," *Journal of Law, Economics, and Organization* 4 (Spring 1988): 95–117; and H. Shelanski, "A Survey of Empirical Research on Transactions Cost Economics," unpublished manuscript, University of California, Haas School of Business, Berkeley, CA.

activities can be cost-effectively outsourced in these two environments. To get insights that can be applied to DoD in light of these important differences, we explore strictly commercial experience.

The empirical literature on commercial outsourcing decisions points to six key factors—factors that, fortunately, are consistent with common sense. Commercial firms prefer not to outsource activities (1) that require real-time control of a complex process, (2) whose production depends heavily on joint use of customized assets (defined below), or (3) when the product of the activity cannot be specified well enough to allow the buyer and seller to write an enforceable contract. Even if a firm decides to outsource an activity after making the above considerations, (4) it may keep some portion of the workload or a closely related workload in-house to train the managers who will oversee the outsourced workload and to maintain knowledge that can be used to judge the performance of that workload. On the other hand, even if the considerations above suggest that the firm should keep an activity in-house, the firm may outsource at least a portion of the workload (5) to get access to technical or market information that the firm cannot get in any other way. Whether a firm prefers, on the basis of the factors above, organic or outsourced production, (6) it may decide to split the workload between in-house and external sources to impose discipline on both by exposing both to additional competition. This chapter considers these six factors, then reviews briefly what they tell us about the government-contracting practices that should be used to outsource defense-support services.

REAL-TIME CONTROL AND COORDINATION

Firms worried about maintaining real-time control over certain assets tend to keep those assets under their direct control. Note that this concern—voiced so often by military commanders who insist on real-time control in combat—exists in an environment totally unrelated to military conflict. This insight probably has implications in parts of DoD not typically associated directly with combat.

Commercial firms worry that the contractual boundary weakens their real-time control. They worry about a variety of opportunities for misunderstandings or disconnects in effective information flow in real-time:

- Incentives and corporate cultures differ in different organizations. These differences raise the potential for misunderstanding and diminish a buyer's ability to get exactly what she wants when she wants it.
- Differences in information systems limit connectivity between a contractor and the DoD, just as they do between different armed services in DoD.
- Suppliers who use secrecy to protect their intellectual property rights in their processes limit buyers' knowledge about these processes and, hence, about what they are actually buying or could buy, given more-complete information.

All these factors lead to differences in expectations during planning and execution.

Effective longer-term contracts give a commercial buyer and seller a safe place in which to nurture mutual trust—something that takes a long time to develop. As that trust grows, a buyer and seller open their organizations and processes more and more to one another, allowing better connectivity, better understanding of how to match requirements and capabilities, and, ultimately, better ability to resolve differences as they face inevitable surprises in their relationship over time.[3] The "just-in-time" material management initiatives coordinated between firms are the most visible recent product of such efforts at integration.[4] More-effective information systems support all such efforts. But commercial buyers and sellers could not exploit such information systems without improved long-term contracts to help them share the gains from using such improved information systems to advantage.

---

[3]A *surprise* is something unexpected and, hence, something the parties have no explicit plans for managing.

[4]Under "just-in-time" inventory management, a part appears where it is needed for incorporation in a higher-level assembly only when it *is* needed. Taken to the extreme, this management concept eliminates the need for working inventory. Users of this concept try to get as close to this ideal as is cost-effective.

JOINT USE OF CUSTOMIZED ASSETS

Customized assets are assets that are more valuable to a buyer and a seller when they use them together than when either attempts to use them with anyone else. The classic example is a set of tools and dies that only one stamping plant can use to create auto parts for only one auto manufacturer. Over time, auto manufacturers have tended to keep such stamping plants in-house. Similarly, when a firm writes software that works only on VAX hardware, the firm is inevitably tied to the VAX hardware manufacturer and support services, for the life of the software and its upgrades. Concern about such interdependence, or jointness, has encouraged commercial firms to use software written in languages compatible with UNIX and DOS, which many hardware manufacturers support. In general, successful commercial firms try to avoid such interdependence.

The special relationship created by such customization creates a surplus value that the buyer and seller are both tempted to extract for their own, opportunistic use. In the limit, such opportunistic behavior can generate enough conflict to destroy the full value of the surplus. In the commercial sector, successful firms try to protect such surplus value by avoiding such relationships.

If a buyer asks an external supplier to use customized tools and dies, the buyer is likely to retain ownership of the tools and dies and design them so that more than one supplier can use them to stamp parts. In a DoD context, the Services often pay to have test equipment designed for use with only one weapon system. To emulate commercial practice, if they were to outsource support for such a weapon system, the Services would retain ownership of the test equipment and provide it to the support firm, whoever that was, to use as part of its contract. This option is viable only if the Services have the legal right to do so. Typically, when a Service contracts for the design of a test stand, it buys the right to use the test stand in an organic support facility, but not in a contract setting. This limits DoD's ability to emulate commercial practice in its support of many existing weapon systems.

A successful commercial buyer will typically prefer a generic standard like that offered by UNIX, even if a VAX operating system offers some advantages that UNIX does not. The buyer gives up small advantages offered by customization to maintain access to multiple suppliers.

Only if customization offers large advantages do buyer and seller turn to it. When they do, they seek to define a long-term relationship typically referred to today as a *strategic alliance.* Such alliances are hard to exploit cost-effectively. A recent McKinsey study of over 200 alliances found that the key to success comes from coordinating the activities of two organizations with strongly complementary skills.[5] If either partner learns enough about the other's skills to do the work of the partnership alone, the advantage of a partnership is drawn into question. The difficulties of managing customized assets across a contractual boundary then dominate the advantages of coordinating complementary skills through an alliance, and one partner buys the other out. If DoD were the buyer, this would mean that DoD had brought the activity in question back in-house—which is exactly what happened within seven years to about half the alliances in the McKinsey sample.

The lesson for DoD is a paradox: The more control DoD exercises over contractors to protect its investment in the customized assets its contractors use to provide sophisticated support services to DoD, the harder it is for those contractors to provide the benefits typically attributed to commercial practice. In this regard, it is worth noting that defense contractors have not been a major source of the innovations now being touted in the commercial world—at least in part because DoD's contracting arrangements, which were designed explicitly to help DoD manage one-on-one relationships with specialized providers, have effectively socialized the defense sector of the private economy. This sector looks more like the government in many ways than like the rest of the commercial sector. Under such circumstances, where is the complementarity that justifies the difficulties of managing a strategic alliance?

To justify such difficulties, DoD must find ways to free its contractors without losing effective control of its investments. Until it does, it may well be most cost-effective to protect these investments by keeping sophisticated support services in-house. One way out of this

---

[5]Joel Bleeke and David Ernst, "Is Your Strategic Alliance Really a Sale?" *Harvard Business Review* 73 (January–February 1995): 97–105.

problem is to qualify and maintain dual sources for support services. The presence of the second source dramatically simplifies monitoring and oversight of the first. Sometimes a second source brings down prices offered to DoD enough to justify the additional costs of the second source; sometimes it does not.[6] However, DoD has a long history of dual sourcing that raises doubts about it as a practical strategy.

Commercial innovators have found a way around the problem of sustaining a second source for every purchased product. Toyota, for example, typically maintains a single source of a part, such as a carburetor, for each model of car that it builds, but it maintains different sources of carburetors for different models. It then carefully monitors the performance of each supplier. If one source cannot achieve Toyota's expectations over time, Toyota has a ready source elsewhere among its other existing suppliers. Hence, while seeking long-term relationships with each of its suppliers, Toyota has found a cost-effective way to sustain effective competition among them and has, in fact, replaced 40 percent of its suppliers over a 10-year period, despite its image for seeking continuity with its suppliers. Similar opportunities exist for DoD if it can qualify support suppliers across weapon systems and coordinate its contracting across systems.

Another way out of the problem of realizing the benefits of investment in customized assets is to place higher reliance on generic standards. Such standards broaden the field of available suppliers, relieving the buyer of its need to control the seller. The more suppliers there are in a market, the harder it is for any of them to exploit a buyer—and the easier it is for a buyer to achieve a satisfactory match with a supplier. DoD is currently attempting to broaden its use of generic standards, but it has a long way to go. DoD has a great deal of experience using customized assets with contractors; consequently, it will take time to induce a culture committed to customized military specifications to accept commercial standards. It will take a new approach to contracting to bring contractor performance with customized assets up to the standard DoD routinely expects from its organic support activities. Unfortunately, in the ab-

---

[6] J. J. Anton and D. A. Yao view dual sourcing somewhat differently in "Measuring the Effectiveness of Competition in Defense Procurement: A Survey of the Empirical Literature," *Journal of Policy Analysis and Management* 9 (Spring 1990): 60–79.

sence of the technical data required to create private competition for support services, DoD lacks a tool often available to commercial buyers.

DIFFICULTY SPECIFYING REQUIREMENTS

Commercial firms prefer not to outsource activities with poorly specified requirements for several reasons: It is hard to write requests for proposals for such activities, it is hard to justify a choice between alternative sources, and it is hard to oversee performance once a source is chosen. When these problems exist, commercial firms prefer to keep activities in-house.

To see how this logic works, consider the following distinction. In the background of any relationship between independent, separately owned organizations is the option of going to court to resolve a disagreement. Although organizations usually resolve their differences without going to court, they write contracts with specific language so that they can seek legal redress if necessary; that option affects their relative bargaining power in and out of court. Courts have much more limited jurisdiction over disputes between the divisions of organizations. In the background of any intraorganizational dispute lies the power of fiat that rests in the leadership of that organization. That fiat may have to reflect subtle intraorganizational politics, but it never requires the specificity that courts require to address a contract. For this reason, private firms tend to merge with one another, under a common ownership, when the terms of their relationship are too ambiguous to be enforced in court.

In the context of this discussion on specifications, this insight suggests that, the harder it is to use the courts to enforce a contractual agreement with a private supplier,[7] the better it is to choose an or-

---

[7]The conditions that complicate judicial enforcement grow from inherent uncertainty in the relationship between the buyer and the seller. That is, it is either impossible or not cost-effective to resolve all uncertainties likely to arise over the course of the relationship. If it were, the buyer and seller could write complete contingency clauses into their contract and resolve all uncertainty at the beginning. When such completeness is possible, a discrete transaction can occur; free and open competition is ideally suited to police such a transaction. When buyer and seller must enter a relationship without resolving all uncertainties, the market begins to lose its power. Buyers and sellers often deliberately choose to leave important elements of contracts

ganic supplier. For example, it has been notoriously difficult for DoD to enforce performance-type warranties—warranties that guarantee that a system will realize a certain standard of performance in operation rather than passing an acceptance test at the time of acquisition.[8] This problem arises because so many factors contribute to operational performance that juries have been reluctant to place blame on contractor support services. The problem becomes more difficult when the system uses more-advanced and more-subtle technology. The absence of enforceable warranties makes outsourcing less attractive.

It is worth noting that performance-type warranties on commercial systems analogous to DoD systems—for example, high-performance aircraft engines—have often worked well when offered to commercial firms. Two explanations have been advanced to explain the difference with DoD experience. One is that DoD uses higher technology in a less predictable and less forgiving environment: It is simply harder to specify the fault trees associated with DoD systems than those associated with more-mature commercial systems operated in a more routine fashion.[9] That is, DoD often uses systems that do not lend themselves to establishing fault. The other is that supplier firms are less likely to challenge a commercial buyer's claims under a warranty if they expect such claims to hurt their long-term reputations as reliable suppliers. As noted above, reputation

---

open, planning to renegotiate terms in the future as surprises arise. So a successful contract need not be completely specified. But as parties to a contract choose to leave more and more open, they must rely more and more on their mutual goodwill, tempting fate to offer surprises where their interests differ markedly. The conditions that complicate judicial enforcement, of course, complicate efforts to manage an activity organically as well. Empirical evidence on commercial firms suggests that judicial enforcement tends to suffer more than organic management, thereby raising the relative advantage of organic production. For an extended discussion of these issues, see Macneil, 1980.

[8]For an excellent discussion of DoD experience with warranties, see Robert E. Kuenne et al., *Warranties in Weapon System Procurement: Theory and Practice,* Westview, Boulder, CO, 1988.

[9]A *fault tree* is a list of all the things that can go wrong with a system and how those things are related. It explains how one failure can spawn another. Hence, the list allows a diagnostician to trace an observed failure to its ultimate source, just as a genealogist can trace someone's blue eyes to an ancestor with blue eyes. A fault tree is called a *tree* because, like a family tree, it looks like a tree. Analysts routinely use fault trees to price warranties—to determine how much someone selling a warranty must ask for it to break even.

plays a smaller role in the current defense-contracting environment.[10] Until that environment changes, DoD outsourcing will not be able to tolerate the same degree of vagueness in contract specifications that commercial outsourcing does today.

Analogous problems arise between DoD buyers and DoD sellers. The Navy, for example, might be reluctant to contract with a defense agency or another armed service for a support service if it could not state clearly what it wanted in the contract and how it would judge performance in practice. The specificity required by a court is not at issue here, because one part of DoD cannot sue another for failure to perform under a contract. But enforcement of any agreement becomes more difficult as the organizational distance between DoD buyer and seller increases, thereby forcing the problem to higher levels in the Department of Defense. The harder it is for a buyer to specify what it wants, the more it wants to avoid a dispute that will be difficult to resolve, and, consequently, the closer the buyer wants the seller to be within DoD.

Of course, buyers often overstate the difficulty of specifying what they want and how they will enforce an agreement, contractual or otherwise.[11] Being specific can force difficult questions about goals and trade-offs to the surface. Specific performance requirements can also increase accountability and, with it, the opportunity for conflict inside organizations. Divisions within DoD often specify far less than they could in their intra-DoD memoranda of agreement. When this occurs, contracting for services—or at least forcing a formal source selection or rigorous administrative procedure to make contracting an option—could cost-effectively force DoD buyers to be more specific about what they want.

Forcing the deliberations and decisions required to write specifications could be one of the significant contributions of contracting for support services: We should not have to take at face value claims

---

[10]Knowledgeable observers have noted that General Electric and Pratt and Whitney tend to be more responsive to commercial airlines using their engines than to the armed services, which use very similar engines also made by them—often engines with almost identical cores and other subassemblies. One explanation is that reputation plays a larger role in commercial contracting than in DoD contracting.

[11]That is, buyer and seller could cost-effectively increase the specificity of their agreements.

that what a buyer needs is too subtle to write an enforceable contract for it. This may simply be an excuse that contracting can flush out.

DEVELOPING KNOWLEDGE TO OVERSEE OUTSOURCED WORKLOAD

Even when successful commercial firms decide that it makes sense to outsource an activity, they recognize that they cannot outsource the *oversight* of that activity. They need enough knowledge of the activity in-house to provide satisfactory oversight; in fact, that knowledge about its suppliers, cumulated and integrated across all of the processes needed to service the firm's customers, may be the principal value added—the core competency—that justifies a firm's existence.

To preserve that knowledge and allow it to grow as circumstances change, commercial firms often retain some workload in-house, even as they outsource most of it. The in-house workload then acts as a training base that preserves the firm's ability to work productively with its contract suppliers. Direct observation of the performance of the in-house activity also provides a yardstick that the buyer can use to measure the performance of a contractor; such a yardstick is most helpful when competition among sellers is limited. More generally, the more complex the continuing buyer-seller relationship is, the more important such in-house knowledge becomes.

An example of this practice in DoD is the Air Force Air Education and Training Command's decision to contract for support services at all of its bases but one. That one in-house base provides oversight talent for all the rest. Such in-house capability is especially important to an organization such as DoD, which limits lateral entry of personnel.

ACCESS TO EXTERNAL INFORMATION

A firm that has traditionally conducted functions such as the management of facilities, materials, or data in-house can easily lose sight of changes taking place in or relating to those functions outside the firm. In most firms, such functions do not draw persistent, focused attention from senior management, which is more concerned with the core activities of the firm. In fact, managers of cost centers providing such services within a firm can easily become complacent about their lack of visibility inside the firm and can grow increasingly threatened by innovation outside the firm that might disturb their comfortable positions.

Outsourcing such functions/activities forces a firm to remain in touch with external developments. At a minimum, a buying firm deals with a provider who must keep track of technological and market developments to remain competitive in its main line of work. As long as the firm is visible as a buyer of services, other potential providers approach the firm and offer new services to get a foothold with the buying firm. These competitors bring the buyer information about its opportunities, even if the buyer does not know how to ask for it.

A more proactive buyer actively scans the external environment to find the best provider available. And even once a proactive buyer establishes a relationship with a provider, it seeks to learn more about the provider's capabilities to improve its match with the firm. This interaction transfers external information into the buying firm that it cannot acquire in any other way.

All these opportunities for getting information about external options grow in value as the outside environment becomes more dynamic. The information gained in any of these modes can bring value to the buyer that dominates the benefits the buyer might expect from exclusive organic production. But a buyer need not outsource all of an activity to get the information it values from external sources. Splitting its workload between in-house and external sources can allow the firm to benefit from the higher cost-effectiveness of its in-house source and the strategic value of the information it garners from the external market. One potential use of that external information may be as a yardstick for judging the in-house source and ensuring that the in-house source performs as expected against the external alternatives.

DIRECT, CONTINUING COMPETITION BETWEEN CONTRACT AND ORGANIC SOURCES

It is one thing to split workload between internal and external sources to get the benefits suggested above; it is quite another to place internal and external sources in direct, continuing competition with one another and to use that competition to determine how to split workload between them. Commercial firms have often used internal sources in this way to discipline their external sources, and vice versa.

This is a tricky game and often fails. Because it presents some ambiguity about how the firm sees its boundaries, it is sometimes difficult to reconcile with a firm's strategic view of its future. More tangibly, it invariably raises questions of a level playing field, often because a firm does favor one source or the other on the basis of other arguments presented here. If thc firm can keep both internal and external suppliers satisfied enough to remain in competition with one another, however, this approach offers a special form of outsourcing.

Such outsourcing will be most attractive to DoD when its only source of supply is a single, private-sector firm—typically, the original equipment manufacturer (OEM) of an item to be supported—and an organic source. This situation occurs when DoD has bought the technical data and rights to create an organic capability but the OEM retains the sole right to provide technical data to private-sector suppliers. When it does occur and the level of demand for a service is high enough to justify two sources of supply, it may be preferable for DoD to use both the government source *and* the private-sector source rather than one of them exclusively.

IMPLICATIONS FOR CONTRACTING EFFORT

This discussion inevitably brings us back to Figure 1.3. It should be easier to understand now that the deeply held goals that underlie DoD contracting policy are likely to have unintended negative effects. As noted in Figure 3.1, these goals are likely to inhibit DoD's ability to establish long-term relationships with a small number of high-quality suppliers, to adopt source-selection criteria that could help DoD choose more cost-effective sources of supply, to manage fraud and abuse in a simpler manner, and to achieve optimal rather than merely acceptable performance.

These problems clearly do not prevent DoD from expanding its outsourcing of support services or from improving the methods it currently uses to contract with suppliers. But they do limit DoD's ability

RAND *MR734-3.1*

	Central Goal of Government Contracting	Unintended Effect of Government Goal
Equity	• Ensure equal access for *all* qualified suppliers	• Discourages long-term relationships and deep investment in suppliers • Limits source-selection criteria
Integrity	• Prevent *all* government-employee and -contractor fraud	• Imposes greater costs than it avoids • Limits discretion and buyer-seller matching
Efficiency	• Get *lowest cost* for acceptable performance	• Deemphasizes performance • Complicates efforts to induce continuous improvement

Figure 3.1—Three Central Goals of Defense Contracting Discourage Use of New Commercial Practice

to emulate commercial success with outsourcing, and they force DoD policymakers to think carefully about the contract vehicles available as DoD does expand its outsourcing of support services. To the extent that DoD can improve its approach to contracting, expanded outsourcing will be easier.

Experience in the commercial sector over the past 15 years makes three points clear:

(1) If DoD can increase its emphasis on performance relative to cost in defining criteria for source selection, contractor responsiveness will improve. Current federal contracting regulations allow this today. Now the contracting culture within DoD must adjust for such responsiveness to occur. The contracting community must develop the new language that will reflect an increased concern for performance in source selections and contracts. By its very nature, *performance* is harder to describe in objective language than the elements of price and cost that dominated contracting for support services in the past. Abandoning tested language guarantees litigation to test the new language. The contracting community will bear the costs of

this litigation and the risks it creates more directly than anyone else. Because DoD has created a body of standards and incentives over the years that encourages contracting officers to be conservative, this increased uncertainty will be threatening to them.

(2) Greater emphasis on reputation and generic standards will simplify contracting and give DoD access to commercial firms that choose not to provide services to the federal government today. Time will tell whether DoD efforts to rely more heavily on reputation can stand up to legal challenge. Again, such efforts will take time and will impose increased uncertainty on a risk-averse contracting community. No regulations discourage DoD from using generic standards; but the current DoD culture does. After decades of developing technical standards that often spread beyond DoD to become commercial standards as high technology diffused from DoD into the commercial economy, those designing new systems in DoD will have great difficulty accepting the passive role of approving standards, and the technology associated with them, that have been created in the commercial world.

(3) Most important, long-term relationships can allow better integration between DoD operators and the contractors that support them. Experience in the commercial sector tells us that it takes time to build the mutual trust that allows such integration to yield satisfying results. Current DoD regulations do discourage such relationships in the absence of a sole source and make it hard for DoD to maintain small numbers of sources, each of which works closely with DoD. Contractual innovation in this area offers the greatest opportunity for benefits to DoD over the long run.

SUMMARY

Successful commercial firms become more reluctant to outsource an activity as (1) real-time control of a complex process becomes more important, especially in an uncertain operating environment; (2) the potential joint value to buyer and seller of employing customized assets grows, especially in an uncertain operating environment; or (3) it becomes harder to specify the performance desired in a contract well enough to enforce the contract in court.

Successful commercial firms are more likely to split any particular workload between internal and external sources when the following circumstances apply: (1) an external source looks more cost-effective, but an internal source provides a setting for training personnel who will oversee the external source as a yardstick against which to compare the performance of the external source; (2) an internal source looks more cost-effective, but an external source provides a source of market and technological information that the buyer can acquire in no other way; or (3) having an internal source and an external source directly compete against one another on a continuing basis increases the competition that each faces, thereby enabling the buyer to extract better performance from both.

Although DoD's institutional setting differs from that of most successful commercial firms, the same factors that promote the decisions in the commercial sector discussed in the preceding paragraph relate to decisionmaking in DoD as well. Innovations in commercial contracting practices over the past 15 years have encouraged commercial firms to rely more heavily on external sources of support services by making it easier for these firms to overcome concerns like those raised above. DoD should recognize that its ability to outsource cost-effectively depends heavily on the contracting vehicles that it can use. Current DoD contracting practice severely limits DoD's ability to follow the commercial move toward increased outsourcing. Contracting reform could help DoD overcome a number of important barriers to expanded outsourcing.

Chapter Four

IMPLEMENTATION

By applying insights from the last two chapters, let us suppose that activities that DoD could outsource in a way that satisfies key decisionmakers and results in cost-effective outcomes for DoD have been identified. What should DoD do to ensure that it successfully outsources such activities? To examine this question, we reviewed RAND's work over the past 25 years on the implementation of major organizational change in large federal agencies.[1] We also reviewed the current business literature on how to induce change in large commercial organizations.[2] Happily, the two reviews yielded compatible results. DoD will have to change the way it thinks about contracting for support services in fairly basic ways. This chapter presents our findings.

BASIC NATURE OF THE PROBLEM

Large, successful organizations typically institutionalize and thereby preserve the successful values and procedures that define the status quo. DoD is no exception. Where organic supply exists, DoD organizations will resist any large change, no matter how desirable.

---

[1]A useful overview appears in Gail L. Zellman et al., "Implementing Policy Change in Large Organizations," in National Defense Research Institute, *Sexual Orientation and U.S. Military Personnel Policy: Options and Assessment,* MR-323-OSD, RAND, Santa Monica, CA, 1993.

[2]A useful overview appears in Arnold Levine and Jeffrey Luck, *The New Management Paradigm: A Review of Principles and Practices,* MR-458-AF, RAND, Santa Monica, CA, 1994.

To understand this behavior, consider the problem of coordinating thousands of people to plan for and execute a single task. Large organizations attack this problem by creating a robust structure for that task—a structure that includes standard operating procedures for handling routine aspects of the task and a set of values, reporting relationships, personal relationships, and more-or-less well-defined incentives for handling aspects that are not so well-defined. Large organizations survive and prosper relative to alternative institutional forms because their structure works well for many tasks. It works well because people do not have to think about many aspects of how to get the task done; they do what they have done in the past. Trying to change such a structure is always hard. It is hardest in large organizations because so much of the structure is informal and takes on a life of its own, a life that can become fairly distant from the interests of top-level leadership. Change agents have no easy mechanism for identifying and dismantling such informal structures. If DoD cannot find a way to overcome these established patterns of practice, any effort to expand outsourcing significantly will fail.

Many observers question this blunt statement. For them, expanded outsourcing is simply about seeking an alternative source of supply and affects nothing more than the procedures a purchasing agent uses to qualify suppliers. But commercial firms that have aggressively reevaluated which work to do in-house and which to contract for have faced a profound cultural shift. Their interest in the details of production changes from encompassing all the elements of their products to focusing on the key elements that they can produce better than anyone else. For the remaining elements, they now seek ways to get what they need from others who are beyond their direct control but are better able to decide how to provide what is needed. Management emphasis shifts from the details of execution to oversight based on performance expectations. The very nature of command and control changes as partnerships replace clear hierarchies. Long-term relationships govern these partnerships. They are relationships based on mutual trust and are disciplined by a common concern about reputation and by the availability of alternative sources and customers if expectations are not realized. Such a new world threatens those in an organization who understand how the old world works and how to benefit from it.

Changing all the processes discussed in the last two chapters could ultimately bring a similar degree of cultural change to DoD. A quick summary of processes that would change includes the following:

- Definition of requirements, initially in terms of the goals of outsourcing, then in terms of the support concept, and ultimately in terms of the technologies in weapon systems that will require life-time support
- Source selection, including who participates in the source selection, how suppliers are qualified, and which criteria the process applies for choosing among sources
- Contract design, including emphasis on a different set of criteria, much closer attention to specific risk mitigation, and much greater attention to mechanisms that can make continuous improvement attractive to both the buyer and the seller
- Contract oversight that allows real-time control tailored to the needs of the military user and ensures that planning for risk mitigation works as risks manifest themselves over the course of a contract
- Integrated planning, training, and execution to ensure that contractors are as well tuned into the peacetime and wartime needs of military users as are organic support activities.

Outsourcing can expand without making all these changes. But the real gains likely to come from effective outsourcing—from exploiting the core competencies of commercial firms that specialize in providing the support services that DoD wants—will come from *reorienting* DoD so that it can consume these services effectively. Such reorientation will ultimately make *all* these changes valid issues for consideration. And as DoD makes more and more of these changes, the basic nature of its culture will shift. Conversely, unless the basic way DoD thinks about planning for and executing support services shifts, changes as extensive as these will not be sustainable.

CHANGING LARGE ORGANIZATIONS

Inducing significant change in a large organization requires high-level commitment to a long-term implementation plan.

Experience shows that such change cannot occur without visible and persistent support from the top leadership of the organization. The senior leadership must show its aggressive commitment at the opening of any effort to change: It must clarify what the broad organizational goals are that will guide the effort, and it must define the basic institutional arrangements that the organization will use to manage the change. Central to such arrangements is a champion who has clear authority to act day-to-day in the name of the senior leadership and to work across traditional organizational and functional boundaries to seek new connections and information flows inside the organization. Large-scale change is unpredictable by its very nature. As broad policy guidance works its way into the heart of the organization, the specialized, but localized, expertise of many people will be brought to bear. Full implementation cannot occur without that expertise, but it must also ensure that local expertise remains responsive to the broad initiative. The champion must be prepared to ensure that, as the initiative unfolds, it continues to serve the broad organizational goals that top management has identified.

As this process proceeds, the senior leadership must retain oversight of the process. It can do so most easily by ensuring that it itself reviews progress at regular intervals and either reaffirms its continuing support or indicates how it wants the process to change. Without this continuing involvement, the inertial forces of any large organization will gradually but surely overwhelm any effort to break the accepted approach to business.

Large-scale change takes time. DoD leadership needs to be prepared to make the commitment required to see such change through. And the normal turnover of political appointees and military personnel in DoD will make any meaningful commitment to a long-term program difficult to sustain.[3] In particular, no individual leader or team of leaders will be able to see the program through to successful completion—a successful completion that will be impossible to achieve without a formal implementation plan that phases in change. The plan must specify an institutional way to identify candidates for out-

---

[3]More continuity is better all around, but the position of champion may be unstable as well—especially if he or she is a military officer. The civilian deputy to the commander may be the person to promote continuity—across the board.

sourcing and screen them. To guide that screening process, it should also provide substantive suggestions on which types of activities to consider first.

To promote and sustain expanded outsourcing, the leadership will probably, at a minimum, have to create and maintain effective "walls" between those parts of each Service that buy and sell support services. Such walls were provided in the early 1990s by regulations that governed public-private depot-level maintenance-source selections.[4] Even though such a division is not required in the absence of formal source selection, it helps reorient values and priorities, making officials who do the buying more concerned about getting best value and less concerned about protecting the jobs of suppliers. Such a change touches deep-seated values.[5] Commercial firms seeking similar types of cultural change have often had to remove and replace many senior and middle managers to eliminate the latent processes and values supporting the status quo.

OSD policy currently prohibits the Services from using formal source selection to compare organic and contract support alternatives. This prohibition is likely to continue until the Services can develop accounting systems and oversight arrangements that ensure a level playing field for all offerors, public and private. In the meantime, the Services will need to develop administrative mechanisms to compare public and private alternatives. At least two general approaches are possible.

The first uses broad evidence from commercial practice and other sources to ask whether an organic or a contract source is more likely to offer the better performance relative to the goals of the senior leadership of DoD. The discussion in the preceding two chapters suggests that the following broad types of activities are most likely to be attractive outsourcing candidates:

---

[4]In-house shops offering to supply services to DoD combat organizations were subject to the same restrictions that private offerors faced. In particular, their access to data and ability to communicate with buyers *ex camera* were restricted.

[5]Such change can, of course, damage the effectiveness of direct command-and-control links within DoD. Decisionmakers should weigh the cost of this damage as they design an implementation plan for outsourcing.

- Activities without a direct effect on combat outcomes
- Activities that commercial firms often outsource
- Activities that many commercial firms currently provide to other organizations.

Combat activities are strategically critical to DoD's overall performance and place a premium on real-time control; commercial firms tend to keep such activities in-house. Where outsourcing is common in the commercial sector—often including a wide variety of generic business services, such as generic information-management systems; payroll, financing, and accounting systems; facilities management; and so on—DoD is more likely to find multiple high-quality offerors who can compete for the DoD workload. DoD is also more likely to find commercial source-selection and contracting exemplars that it can attempt to emulate as it expands its own outsourcing.

Cities routinely contract for many of the services associated with base operating support. Housing has many close analogs among commercial firms that operate rental properties and other residence facilities under franchise contracts. Many simpler maintenance tasks, such as cleaning and painting, should also be easy to outsource.[6]

The second general approach uses a side-by-side comparison of in-house and outsourced alternatives and assesses their relative performance. Formal source selection could serve this purpose if OSD allowed it. In its absence, the Services can use an administrative approach that collects similar information but does not impose prices and other terms determined by the source-selection process or allow appeals when the parties object to the results. A side-by-side comparison would presumably consider the following issues:[7]

---

[6]Cf. Clay-Mendez, 1995, which provides useful guidelines and a more detailed inventory of activities that could be outsourced without complex contractual innovation.

[7]I give special thanks to Carl Dahlman, Ken Girardini, and Nancy Moore for many fruitful discussions about these factors. Future RAND work will elaborate on them.

- Is outsourcing this activity compatible with DoD's basic strategy for providing defense services?
- Is outsourcing feasible? (For example, do suitable commercial sources exist, are they interested in serving the government, and can the government write an enforceable contract?)
- Can all key risks be satisfactorily mitigated? (For example, can DoD ensure access to the service when and where it is needed, in the form requested?)
- Given actions taken to get satisfactory answers to the questions above, how do the alternatives compare relative to DoD's basic goals?

Note that, to answer these questions, the buyer must consider not just the nature of the activities themselves but also the likely nature of the relationship between buyer and seller and, in particular, the terms of the likely (as opposed to the desired) contract.

The activities that this second general approach identifies for outsourcing will not be known with certainty until DoD applies the approach. It would be very surprising, however, if—at least in broad outline—it yielded a different list from that generated by the first general approach. That observation raises an important question. The second approach is likely to require far more information and analytic resources than the first. If both yield similar answers, why use the more costly, second approach? The answer is that, the greater the uncertainty about whether an organic source dominates a contract source *and* the greater the resources at risk if DoD makes the wrong choice, the easier it is to justify this more arduous approach.

DoD might consider using the first approach to triage candidates for outsourcing: (1) Strong candidates identified using the first general approach for organic provision and small activities for which there is an organic preference would remain in-house. (2) Strong candidates for outsourcing and small activities deemed doable by a contract source would move immediately to external provision. (3) And candidates with substantial resources in question and high uncertainty about which approach was better would become candidates for applying the second approach.[8]

Whatever institutional arrangement DoD uses to identify opportunities for outsourcing, it must recognize that it should not attempt to outsource all potential candidates at once. Its implementation plan should include provisions for gradual and cumulative movement toward expanded outsourcing. Such an approach offers several advantages. It allows DoD to start with the easiest candidates, which do not require large changes in DoD policies. Their success will build a base that managers throughout DoD can use to grow more comfortable with relying on external sources. As they grow more comfortable, the managers will find it easier to make the policy changes that facilitate further successful outsourcing. Incremental expansion of the portion of DoD support provided by commercial sources allows incremental confidence-building.

Incremental expansion also reduces the probability of serious failures, which could undermine confidence in the program and prevent its long-term success. DoD should anticipate that some efforts to outsource specific activities will fail. Each outsourcing initiative should plan against potential modes of failure[9] and allow a graceful recovery when failure occurs. Such planning deserves special attention early in the implementation program because (1) early failures that seriously degrade the performance of the force pose the greatest threats to the overall outsourcing program, and (2) failure modes will be easier to plan against as DoD gains more experience with outsourcing a particular activity.

---

[8]A parallel exists in recent discussions of product-liability policy. As DoD weighs these alternatives, it would benefit from reviewing analogous arguments, pro and con, about the choice between individual formal source selections and broad administrative determinations. Using common law to resolve product-liability issues one case at a time guarantees closer attention to the details of each case, but it imposes substantial costs and delays and still does not guarantee the most cost-effective outcome in each case because of biases inherent in the common law. Broad statutory solutions are blunter instruments and more susceptible to the discretion of the decisionmakers crafting them, but they may, in the end, be more cost-effective for certain aspects of product-liability policy. A good overview of these issues appears in Peter H. Schuck, ed., *Tort Law and the Public Interest*, W. W. Norton, New York, 1991.

[9]Modes of failure are as various as the activities to be examined. Examples are failure of a data system, a strike, and a facility disabled by chemical weapons.

Throughout this process, DoD should give careful attention to the way its contracts affect its ability to build relationships with contractors. It should seek new ways to contract, and it should plan to expand its privatization efforts as new contracting vehicles open new opportunities.

More generally, as DoD gains experience with expanded contracting and its contracting and requirements cultures begin to adjust to a more commercial orientation, DoD should consider tougher and tougher activities as candidates for outsourcing—those activities that, by the six criteria at the beginning of Chapter Three, look more and more as though they belong in-house. As DoD finds it harder and harder to identify additional candidates that its policies would allow it to outsource with confidence, the implementation of expanded outsourcing will draw to a natural close.

SUMMARY

To expand outsourcing in a way that gives DoD full access to the core competencies for support services being displayed in the commercial sector, DoD will have to change many of the processes it uses to plan for and execute support services. To change all these processes, DoD will have to accept a degree of cultural change that most large organizations find very hard to accept.

Significant change is more likely to succeed in DoD if DoD develops an effective, long-term implementation plan for phased expansion of outsourcing. The plan starts at the top, with clear and sustained support from the senior leadership. It includes the identification of a champion empowered to work across functions and organizations in DoD and to sustain the intent of senior management as implementation moves from defining policy to changing specific practices at depots and installations. It includes a mechanism for ranking activities in terms of the likelihood that outsourcing them would improve the cost-effectiveness of DoD's support services without imposing any unacceptable risks should outsourcing not work as anticipated. And it includes a plan to outsource the most-promising activities first. The plan should give close attention to achieving a series of early successes that build confidence in expanded outsourcing within DoD and enable DoD to outsource more and more-challenging services as its confidence and capability grow over time.

Any outsourcing decision should give close attention to available contracting vehicles that would give DoD access to an external source. The long-term implementation plan should promote contracting reform and build on it, incrementally expanding outsourcing as better contracting vehicles become available.

Chapter Five

RECOMMENDATIONS

This report makes four basic recommendations.

1. **Plan formally for the major organization changes required to implement and sustain privatization.** Privatization involves far more than a simple shift from an in-house source of supply to a contract source of supply. Successful privatization on the scale that DoD is now discussing will require significant changes in many DoD processes. A number of these changes are cultural; they will be far harder to effect and diffuse than just making simple changes in legislation, regulations, or formal policies. For these changes to succeed, DoD needs to develop a workable implementation plan and stick with it over the period of time it will take to achieve these changes.

2. **Focus on improving contracts.** By definition, DoD cannot expand privatization without increasing its dependence on contracts. Effective contractual vehicles hold the key to getting the performance that DoD needs from new contractor sources of supply. Keeping this fact in mind, DoD should carefully coordinate any privatization effort with its efforts to reform acquisition or it should initiate new efforts to improve the contractual vehicles it will use to access contractor sources of supply.

3. **Start with the best candidates for privatization.** DoD should seek early successes in expanded privatization and use those successes to increase confidence in the Congress and throughout DoD that still-further privatization is promising. Conversely, it should assiduously avoid early failures that could discredit the longer-term program of privatization. Both considerations suggest that DoD should start where expanded privatization is easiest to achieve and move—over

time, as contractual improvements and accumulated experience build DoD's capabilities—to privatize ever-more-challenging activities.

4. **Where necessary, protect constituents that privatization might hurt.** DoD should recognize that not everyone will benefit from privatization, even if it is outstandingly beneficial for DoD as a whole. DoD should expect political opposition and seek constructive ways to address it. One approach would seek simple ways to limit the losses induced by privatization. Another would use an extrapolitical process, such as the current BRAC process, to choose a group of candidates for privatization, thereby diffusing the losses caused by privatization.

REFERENCES

Anton, J. J., and D. A. Yao, "Measuring the Effectiveness of Competition in Defense Procurement: A Survey of the Empirical Literature," *Journal of Policy Analysis and Management* 9 (Spring 1990): 60–79.

Bleeke, Joel, and David Ernst, "Is Your Strategic Alliance Really a Sale?" *Harvard Business Review* 73 (January–February 1995): 97–105.

Camm, Frank, *DoD Should Maintain Both Organic and Contract Sources for Depot-Level Logistics Services*, RAND, IP-111, Santa Monica, CA, 1993.

Chandler, Alfred D., *Strategy and Structure*, MIT Press, Cambridge, MA, 1962.

Clay-Mendez, Deborah, *Public and Private Roles in Maintaining Military Equipment at the Depot Level*, Congressional Budget Office, Washington, DC, 1995.

Commission on Roles and Missions of the Armed Forces, *Directions for Defense*, Washington, DC, May 24, 1995.

Harrigan, Kathryn R., *Strategies for Vertical Integration*, Lexington Books, Lexington, MA, 1983.

Joskow, Paul L., "Asset Specificity and the Structure of Vertical Relationships: Empirical Evidence," *Journal of Law, Economics, and Organization* 4 (Spring 1988): 95–117.

Kelman, Steven, *Procurement and Public Management,* AEI Press, Washington, DC, 1990.

Kettl, Donald, *Sharing Power,* Brookings Institution, Washington, DC, 1993.

Kuenne, Robert E., et al., *Warranties in Weapon System Procurement: Theory and Practice,* Westview, Boulder, CO, 1988.

Levine, Arnold, and Jeffrey Luck, *The New Management Paradigm: A Review of Principles and Practices,* MR-458-AF, RAND, Santa Monica, CA, 1994.

McNaugher, Thomas, *New Weapons, Old Politics,* Brookings Institution, Washington, DC, 1989.

Macneil, Ian R., *The New Social Contract: An Inquiry into Modern Contractual Relations,* Yale University Press, New Haven, CN, 1980.

March, James G., *A Primer on Decision Making: How Decisions Happen,* Free Press, New York, 1994.

Schuck, Peter H., ed., *Tort Law and the Public Interest,* W. W. Norton, New York, 1991.

Shelanski, H., "A Survey of Empirical Research on Transactions Cost Economics," unpublished manuscript, University of California, Haas School of Business, Berkeley, CA.

Smitka, Michael, *Competitive Ties,* Columbia University Press, New York, 1991.

Sumser, Raymond J., and Charles W. Hemingway, *The Emerging Importance of Civilian and Contractor Employees to Army Operations,* Landpower Essay No. 95-4, Association of the United States Army, Institute of Land Warfare, Arlington, VA, 1995.

Thompson, Fred, and L. R. Jones, *Reinventing the Pentagon: How the New Public Management Can Bring Institutional Renewal,* Jossey-Bass, San Francisco, CA, 1994.

Williamson, Oliver E., *The Economic Institutions of Capitalism,* Free Press, New York, 1985.

Wilson, James Q., *Bureaucracy,* Basic Books, New York, 1989.

Zellman, Gail L., et al., "Implementing Policy Change in Large Organizations," in National Defense Research Institute, *Sexual Orientation and U.S. Military Personnel Policy: Options and Assessment,* MR-323-OSD, RAND, Santa Monica, CA, 1993.